From Where We Came

Also by Arthur J. Stewart:

Rough Ascension and Other Poems of Science
Bushido: The Virtues of Rei and Makoto
Circle, Turtle, Ashes
The Ghost in the Word

From Where We Came

poems

Arthur J. Stewart

CELTIC CAT PUBLISHING
KNOXVILLE, TENNESSEE
2015

From Where We Came: poems / Arthur J. Stewart

ISBN: 978-0-9905945-9-8

Celtic Cat Publishing
5111 Green Valley Dr
Knoxville, Tennessee 37914
CelticCatPublishing.com

We look forward to hearing from you. Please send comments about this book to the publisher at the address above. For information about special educational discounts and discounts for bulk purchases, please contact Celtic Cat Publishing.

Manufactured in the United States of America
Cover and interior art by Justin A. Dickerman-Stewart
Library of Congress PCN: 2014959648

For the friends I've not yet made, but hope to, in the worlds of anthropology, archeology, paleontology, geology and geography; and for those who work in chemistry, physics, biology, ecology or history; and indeed for any readers whose curiosity might be fired, even a little, to learn that techniques such as multispectral imaging can reveal new information about the Denisovans or the Martellus Map.

Contents

Preface and Acknowledgments

This book is about gains and losses. We have gains and losses every day. Our individual gains and losses are sometimes tiny—so small they're invisible to everyone except us. And sometimes, on a memorable day, a gain or loss can be explosively large—a jump to elation, or collapse to a heart-wrenchingly deep loss. Gains and losses can be sharp and savory, or soft and fragrant—ten thousand fragile examples tell us this, every day: lilac, coconut oil; a sunset over the Pacific Ocean looking west from the coast of Mexico, a hug, the flit of a swallow at dusk. At the individual level, our gains and losses are in essence local: each one is important to us individually, centered here in our temporary bodies and bustling day-to-day lives. Gains and losses are the things that give us meaning.

But collectively, mankind has gains and losses, too—gains and losses that manifest over thousands of generations and thousands of miles. When we look back with the right tools we can almost see them: a ghostly trail of history—the rise and fall of cities, nation-states, and entire civilizations. Our histories show our gains and losses, written in part in ochre and ash on cave walls, in part in cuneiform on clay tablets, in part through hieroglyphs on stone, and in part on sheepskin and papyrus in ancient scrolls. But increasingly, archeological studies are using fantastic new scientific tools in the study of man—airborne LiDAR, for example, can map the near-invisible traces of walls in ancient cities, and investigators now can extract and analyze incredibly small amounts of DNA salvaged from fossilized bones and teeth, allowing determination of genetic relatedness of ancient bones to current man. Through science, we are learning where to best look for our histories, and we are finding new ways to reveal details about them.

Of great importance, science also has expanded our ability to reliably determine the ages of fossils and ancient artifacts. New radiometric, thermoluminescence and incremental dating methods, for example, supplement the widely used 14C dating method, greatly extending the range of our look-back time. These methods are being applied everywhere—to civilizations that lived along the edges of the Atacama Desert, to the remnants of ancient cities in Jordan, to fossil bones recovered from Australia, Spain, Africa and sinkhole caverns in Florida. Through science, we are busy exploring from where we came.

Yet while science is among mankind's most powerful and important constructs, it is limited in what it can tell us about the world. Science is a bit clunky and whimsical at times, in part because of the people who do it, and in part because good science depends upon reproducibility. Although science works best with multiple lines of evidence, each line of evidence inevitably grows, like a leafy tendril, at a different rate. Through time, even facts can get forgotten or lost; these losses, too, are important. We begin to see things differently, because things change. The voids between the facts may get smaller and smaller as the body of science grows, but these voids are like minuscule vacuums: they fill with other human things—suppositions, presumptions and interpretations, elation, love and fear—and these things, too, give meaning.

In a nutshell, here's the problem: because science works so well most of the time, we begin to trust it, perhaps too much. Look carefully at what humans do now, and what we did, and how the world works now and you can still see the little voids between the facts—the millions and millions of crannies, cavities and holes between the facts populate us like pores populate a block of foam or an aerogel. That's where our histories live, deeply embedded. No scientific brush can ever paint that fine.

Many of the poems in this book focus on our complex origins and derive from archaeological and anthropological facts extracted from peer-reviewed scientific articles in scientific journals. But we must honor this fact, too: all we have, in truth, are our aspirations and histories, and these histories are far more nuanced than scientific facts alone.

From Where We Came is my fifth book designed to help bridge the gap between the worlds of science and the literary arts. The world of science is huge. The realm of poetry, in comparison, is tiny, and the overlap between the worlds of science and poetry is far smaller still. But like voids between facts, such overlap exists, and that's important. This book has benefited from the positive feedback of many bemused scientists who have previewed some of the poems included here. I offer special thanks to each of the many scientists who work hard in the field and in their labs, unraveling the mysteries of our collective past. The cover art and the interior art (7-minute quick sketches) by my son Justin Dickerman-Stewart well capture the book's theme from yet another literary-arts perspective. And Jim Johnston, I appreciate your steady guidance and keen editorial eye. Thank you, all, for your help in making this book better!

This book is in part for scientists, and many of the poems offered here stem from ideas, facts or language borrowed liberally from scientific articles. But this book also is for those who view themselves as readers of good literature, with no special skills or interests in science. The sources of information used in 45 of the 48 poems in this collection are identified succinctly in the notes section near the end of this book, associated with the poem's title and page number. This method for assigning attribution was selected for two reasons. First, it provides a reader a direct portal for gaining additional information about the scientific concept or issue being addressed in the poem, should he or she be willing to invest the effort. Second, my hope is that science teachers at every level—from K12 on—can begin to see that the arts and the sciences in fact share much common ground. The referencing method provides readers at least a glimpse of a possible cross-walk for cross-talk, which teachers can use to help their students understand that science and the arts are merely different manifestations of the same thing. The beauty and truth of one, I think, can be captured as truth and beauty in the other. Now, that said, let's get going.

From Where We Came

Science-Flavored Poetry

It's good to stake out
some territory if it doesn't overlap
too much the territories of others and always
      be prepared to pollinate
your ideas with theirs. So what's

what? Today I reach
for functionality first, attempting to exceed
a Mary-Oliver dictum hard.
Note this: I'm not here to spew

a tanglement of related words or pitch
a transitive verb, or a descriptive-only
plethora of slippery black wet things.

No—instead I give each thing

one firm swirl then bear down
to individual flakes and nuggets
glittering among dark grains; each one

seems a cosmos; each one is

a portal to some new place.

From Where We Came

Up from the Olmecs in Mesoamerica, a tendril
thinning as it crept north—man, woman, child;
and north, from Tanzania and Kenya,
through Sudan and Egypt, north
into Syria, Turkey, Iraq and Iran,
trekking, hunting, reproducing, dying—

an unsteady but steady collective move
through ages: from Australia,
north into Indonesia, Viet Nam and China,
a wavering compass needle
from Africa to Spain, from Spain to France and Germany,
north to Denmark, Sweden, Norway, Finland—how far north

we don't know: yet north, tribes
splintering into smaller tribes
north, west and east, the collective
thin as coiling smoke.

Several million years
working up and out of Africa; thirteen thousand years
or more in Egypt; more than a million years
in China, just fifty thousand years or so
in Australia. Take pause to count

the countervailing flows of people south—
from Siberia across the Bering Strait then along the coast
of Alaska, the Pacific Northwest; and the
bend and wend of people east
and south along the coast into Florida; the back-curl
from Spain to Cuba, the flood of people south,
from Mongolia into China, the Picts

merging with Gaels, great sweeps
of intermingling people
from West Africa to the Caribbean;
Mayans merging with Spaniards in Mexico,
Japanese merging with Hawaiians—and the con-
flicted intrusions of people:
from Japan into China and from
what is North Korea, south; and south,
from England into Australia, New Zealand
and South Africa, from Germany into Argentina—

in short, what are we

to make of such things? It is
human by nature to move out, human
by nature to push forth: writing
on clay tablets or on stone with stone
in hieroglyphics the history of an individual life,

or the lives in our communities: which king
begat which sons, which war
was won by which tribe, an arrow of time shot
from left to right or north to south
or right to left, in
dozens of alphabets, whatever it takes to express
self to the world.

We know this: while thinning out,
curling up and moving
north or east or west or south through time

     inevitably

we gain, we lose

great pieces of ourselves: grand ideas
emerge, settle in, get dislodged and shucked off.

Genes are favored or get thinned out,
the hair, the skin, the color of eyes, the width
of a nose, molecular
factors in the blood:

of this great body, yet what else

comes and goes? We looked
back then from valley cave-lip out and up to stars
millions of years ago to constellations
signaling fate, and now from mountain-top
through telescopes to faint wobbles of light as planets
traverse the faces of their centering fires

yet always
the air, moisture-laden and hot or
raw and cold, going in
to lung and cell and out
dealing with the flood of data in, what new

gyres to interpolate
a honey-flow or burst of scent, the keen
of what we call a gull wheeling
the air and the thick reek of kelp and salt-splashed rock,
marsh and rot, frog-call, raccoon

looking back, a black-masked face as it
thin-fingers the water's edge. Desert
beardtongue, *Penstemon pseudospectabilis,* just one
of many species
bursting to bloom
in this particular case following a wet winter, local

to extreme—why call this species out
here, among millions as an example? Because

it is
like us
just one

of many types: variable, widespread; responding
to environment—the preceding year,
the weather, its nearest neighbors, the night-
time swoop and curl of cool air tumbling

down a canyon, attentive
to the flit of fly and bee, to the
whims of dawn, such
glory in early light this moment, this
small instant, now.

Long, Long Ago

Long,
    so long ago,
from the Permian to the Jurassic, lasted Pangaea,
massive land-mass circling the Tethys Ocean
and through a monotony of time
that great supercontinent began frag-

menting, spinning off

in three long, long phases,
Laurasia, Gondwana, Cimmeria,
Africa, Antarctica and Madagascar, South
America, India, and Australia

causing to open,

like slow petals,
the South Indian Ocean, the North Atlantic,
the Coral Sea, the Tasman Sea,
the Norwegian Sea,

and causing to rise

great mountains:
the Himalayans, the Alps, the Andes, and
yes, even
our little Appalachians.

What more

could we ever want
of this Earth, such great history,
our island home?

The First, Second and Third Worlds According to the Hopi

1

In the First World, the First People
had multiplied and spread over the face of the land
and they were happy

at first
but of course

they began dividing and drawing
away from one another until at last
Sótuknang came in a great wind and told them, do this:
follow a certain cloud by day, a certain star by night
to a certain place
where, when they reached it, they were invited
underground to live with the Ant People,
gathering food in summer for winter.

There it was cool inside when it was hot outside
and warm inside when it was cold outside:
and the people lived peacefully.
And then,
while the First People were safe underground,
the First World was destroyed by fire.

2

After a long while, when the First World had cooled
Sótuknang readied the Second World.

He put water where land had been and land
where water had been; then he thanked

the Ant People for their hospitality to the chosen people
and he welcomed out the chosen people

to silver and blue; he welcomed them south

to the Second World,
to spruce, skunk and eagle and they sang
joyful praises to the Creator.

Yet again,
through time,
gradually through time,

people began trading for things they didn't need
and they began squabbling and fighting
until at last only a few retained
goodness in their heart.

So the Second World, too, needed to be destroyed.

Again, the Ant People kindly secured
a place underground for the chosen people.

Then the twins were called and they abandoned
their northern and southern posts, and the world,
with no one to control it,
teetered off balance, spun
crazily and rolled over twice.
Mountains plunged into seas, seas
and lakes sloshed for miles over land
and as the world spun
through cold and lifeless space

it froze

to solid ice.

3

The Third World was east, its color red:
copper, tobacco, crow and antelope.
The people multiplied quickly, made big cities, countries,
a whole civilization; they got entangled
with their own earthly plans. They made shields
of hides and flew on them,
attacking other cities, making corruption and war.

So once more

Sótuknang had to intervene. He told Spider Woman
to save the people in hollow reeds, with
white cornmeal dough for food; then he
loosed the waters upon the world—waves
higher than mountains, continents sinking
under seas, and horrific rains

and the people
bobbed and floated in the sealed-up reeds

for long time until Spider Woman
brought them out to a little piece of land
near the tip of a tall mountain, sticking up
through water. But it was much too small
to be the Third World. So again the people traveled,
east and north, in reed boats and on reed rafts, landing
at last on a large island rich and flat with trees and plants,
seed-bearers and nut-bearers, and they stayed there

for years, until Spider Woman at last
made them leave again, reporting
        this island
was not the Fourth World.

So again, paddling on reed rafts and in reed boats
they went east and north until
with great mountains the Fourth World
was found: and Sótuknang

greeted them there, saying

Yes, this is

4

unfinished
the Fourth World: it has height and depth;
heat and cold; beauty
and barrenness. It is not easy.
Each of you should follow your own star.
And he then sank the islands they had visited,
thus ending the Third World.

Beginning

Casting back through time, listen
to the soft trickle of history: listen
to the fragments of stories having much
in common. One way or another,

from nothing came seas, and from seas
came land, and almost always
a spirit from above came down
in godly form and through one

of various ways
made manifest, through spit and clay
or fire and violence,
stolen from the bodies of related gods,

the world as we know it:
light, animals, plants and man.

Denisovans

From one cave in Serbia, one tooth.

Later, a small fragment of bone
from the fifth finger of a juvenile.

Then, another tooth,
not from the same stratum;

   and later,

 a toe-bone,
contemporary to the finger bone,
based on the stratum
in which it was found—perhaps
from a Neanderthal-Denisovan hybrid.

Mitochondrial DNA analyses suggest
Denisovans once ranged widely
over much of eastern Asia,

but there's only traces now

  in Australia,
 New Guinea, the Philippines
and in our genes.

To the Time of Long Ago

Restless bits of DNA can drift
throughout the genomes of human brain cells
and when they do, they can create
cognitive diversity: artists
thumb cobalt blue to canvas, writers
miss-match thoughts and words, and words
flutter up from electronic pages
    like blue butterflies
from Nabakovian novels.

So whereas:
broken ion pumps in petunias
reduce acidity in the petals,
which lets them turn blue;

and whereas:
a large blue whale has a heart
the size of a small car;

and whereas:
tungsten and tantalum, which come
from places like Tiger Hill, and which go
into things like smart-phones and cars,
some of which carry
a deep blue sheen;

and whereas:
cobalt blue is more stable but less intense
than Prussian blue (of the non-
musical, non-racist type);

then whereas:
we should know now
(and let it hereafter be known)

everything we do stems back
to blue and we stem back,
long ago, to the Denisovans,
to the time of long ago.

Lesson on Ancestral Aboriginal Australians

The Aboriginal flag is a black bar over a red bar,
a bold yellow circle centered, splitting
the black and the red, like the sun
heating now the earth by day or like the moon
cools it at night: both lights touch
the iron-rust clay.

Traveling south across the Wallace Line
fifty to a hundred thousand years ago
evidence suggests Denisovans became
ancestral Aboriginal Australians.

They took then the form of a great serpent
and they began dreaming as they travelled
the face of the continent, shaping
the landscape, creating new life.

Now the Rainbow Serpent
has more than two dozen wonderful names
and it rises constantly in everywhen:

it-they came, comes or will come
in non-linear time pushing up
ridges and mountains and creating
gorges, river-ways;
it-they inhabit deep water-holes
and create everything—male or female,

androgynous and hermaphroditic;
the it-they are will be or maybe not—
for homework today, look it up:
bunyips, *Diprotodon.*

End of the African Humid Period

Eleven thousand five hundred years ago,
a great swath of North Africa was much wetter
than it is now: things there
began becoming drier

about 5,500 years ago—lakes shrank
to nothing; rivers
went to trickles, forests
went to grassland, grassland
went to desert.

The Egyptians and Sumerians
gathered along the dwindling rivers wondering
what the heck; they learned

to irrigate, they began
intensifying their cultures.
From such efforts, the Great Pyramids.

Ice-Man Ötzi

Ötzi wore a cloak of woven grass and a coat;
a belt, a pair of leggings, a loincloth and shoes,
all made of leather of different skins;
a bearskin cap with a leather chin-strap.

The shoes were waterproof and wide, designed
for walking on snow. They were constructed
of bearskin for the soles, deer-hide for the top panels,
and netting made of tree bark. Soft grass
went around each foot and shoe,
functioning like modern socks.

His coat, belt, leggings and loincloth were constructed
of vertical strips of leather sewn with sinew. His belt
had a pouch sewn to it, and it contained
various useful things: a scraper, a drill, flint-flake,
a bone awl, some dried fungus.

What else? Oh, yes—he had
a fine copper axe with a yew handle,
a flint-bladed knife with an ash handle; a
quiver of 14 arrows, with *Viburnum* and dogwood shafts:
two of these were fletched and tipped with flint; the others

were unfinished. A long-bow, too, was
    unfinished

By analysis of autosomal DNA, Ötzi is
most closely related to southern Europeans,
particularly to populations in Sardinia and Corsica.
He was
at high risk of atherosclerosis and lactose intolerance,
and he contained

the DNA sequence of *Borrelia burgdorferi,* making him
the earliest known human with Lyme disease.

X-ray analysis and a CT scan revealed
an arrowhead lodged in his left shoulder when he died;
a matching small tear occurred on his coat.

The arrow's shaft had been removed before his death.
they found bruises and cuts to the hands, wrists and chest,
and cerebral trauma indicative of a blow to the head.

Data show this, as well:
Ötzi killed two people with the same arrow,
and was able to retrieve the arrow both times;
blood on his coat was from a wounded comrade—
perhaps carried on his back.

Ötzi's posture in death was this:
frozen body, face down, left arm
bent across the chest. So before death,
the Iceman was turned
onto his stomach to remove the arrow shaft.

Years later, controversy slips up a snaky head:
who should own Ötzi's body? And who
among us carries his lineage today, making them worthy
of seeking, like Ötzi, the northern pole star?
Who, among us, will sing for his family's dead?

Like and Unlike

Unlike
what you might suppose, *Australopithecus*
arose in east Africa, as early as 3.6
million years ago. This creature
shared traits with modern apes and man and gave rise
to at least six species, which spread
across the continent before becoming
extinct, two million years ago.

They were
diminutive and gracile:
    they did not

make canoes and paddle their way to Australia.

Perhaps they knuckle-walked now and then,
like an orangutan—
    or maybe not; they were
frugivorous to large extent,
based on tooth-wear.

    Found first

in the place of the lion in South Africa,
the Taung Child: it was classified
as *Australopithecus africanus*: scrutiny revealed it
as the skull belonging to a three-to
three-and-a-half year-old child,
killed by a large bird of prey.

Bluestone Rocks of Stonehenge

The bluestone rocks of Stonehenge did not
come from the Preseli Hills of Pembrokeshire,
a hundred eighty miles
west north-west of the site, in Wales. No,

X-ray analyses show they came instead
from a hill called Carn Goedog, only a mile
from the Stonehenge site. Whooo boy,
that's embarrassing—looking so hard

in the wrong place!
So, let's move on now to the next
item of business: were the bluestones cut
and moved by man or did they break

and get moved
to the site by glaciers?

Yes and No for the Pigs

It's rare
for something to start as a theory and grow up
to become a fact. Yet consider

    Durrington Walls,
    a place of the living—
    and Stonehenge,
    a place of the dead.

(Now, in point of fact, pigs and cows
died like mad at Durrington Walls, so for them, at least,
it was not
a place of the living.)

But:
from dates scratched into a pig bone
and onto an antler pick found at the site,
the monuments at the two places were built
together in time, about 4,600 years ago.

And rituals at both places involved
great feasts—greasy smoke and sparks:
by analysis of the wear patterns on pigs' teeth

    various conclusions have been drawn.

Most of the pigs
died in winter at the place of the living—
a bleak time for sure, but based
on strontium isotope ratios
the pigs were brought there to die
at the place of the living

from all over

Scotland, England and Wales.

Peccary-Chase

While chasing a group of white-
lipped peccaries through forests in central Brazil,
a team of investigators discovered a cave

the walls of which were covered with drawings
made by ancient man—armadillos,
deer, large cats, birds and reptiles,

human-like figures and
geometric symbols, but
oddly, no peccaries.

Closer to Home

Unveiled recently, by LiDAR, look it up:
ghostly remnants of roads and stone walls—
a former agropolis
under a canopy of oak and spruce
around rural New England towns.

The images show
through shadows the history of the site—
fence-lines, the edges
of roads, the places where buildings were,
even the furrows of fields; they show

how man's structures
affect the environment, by shaping
the types of vegetation that develop
hundreds of years later.

It is right to give homage
to the beautiful tools
we now have for looking
back in time: give thanks for the various
molecular clocks, the steady rates of decay
of different isotopes, the delicate annual layering
and the more abrupt event-driven layering

by gravity, water, air and sun
of so many things—
pollen, ash, silt and bits of clam-shells,
the calcified skeletons
of microscopic life.

Easter Island

From Polynesian islands,
in canoes or catamarans,
a group of people came and colonized.
But in time they got squeezed
through bottlenecks of
starvation, war, slave-trade,
then small pox and tuberculosis.

The population on Rapa Nui fell
from a high of about 15,000 to just over 100
desperate souls in just
250 years.

They lived and died and lost a lot; we gained
rapamycin and mystery. Great

stone heads, most now tipped, remain.

Where, When and How

Near Laetoli,
on the western side of Lake Turkana, in the
Kenyan Rift Valley, 3.6 million years ago
or so—imagine

the things that came together
　　just
　　　　by
　　　　　　chance

A volcanic eruption
big enough to flood the area with fine ash
but not
big enough to kill all life;

then rain, just enough

to wet the ash, making it like soft concrete.

And then
　　my god
　　　　three upright walkers

who made the prints. And then,
before another rain, another fall of ash,
snow-soft, sealing history.

Gone Before it Was

Homo heidelbergensis—a big-
brained human ancestor considered
as a pivotal creature during a
murky period of evolution

    and perhaps

the common ancestor of modern humans and
our extinct cousins, the Neanderthals:

Living
perhaps a half-million years or so ago,
and perhaps
a link between *H. erectus* (earlier)
and *H. sapiens* (now).

But things,
    you know,
get complicated.

The species was named
at first from one jaw found
more than a hundred years ago, and later

characterized by other skulls:
massive brow ridges, large
faces, flattened frontal bones and now

with additional analyses the thought is

    perhaps

the thing that was—the whole species
defined by the name we thought up
perhaps was not: perhaps
it is going,
    going,
gone before it was.

Chinchorros

They were sedentary fishing people
dwelling in scattered villages along the coast
of the Atacama Desert,
in what is now northern Chile and southern Peru.

Some lived

in semi-subterranean circular huts.
Seventy-five percent of their diet was seafood
of one type or another—fish, seals,
marine birds, seaweed, shellfish

(delicate mother-of-pearl fishhooks;
hair and fiber lines, reed nets, harpoons).

They made
mummies of their dead
thousands of years before the Egyptians did:

removing the soft tissues, inserting
sticks to reinforce the bones,
stuffing the body with plant material,
covering it
with ash paste,

        fitting the face

with a black clay or a red clay mask
before wrapping the body in reeds.

Natufians

Thirteen thousand seven hundred years ago
in what is now Israel,

Natufians buried their dead in cemeteries
with tortoise-shells and stone tools and

bone tools and
some with dogs as pets, in

mud-lined bedrock-chiseled graves

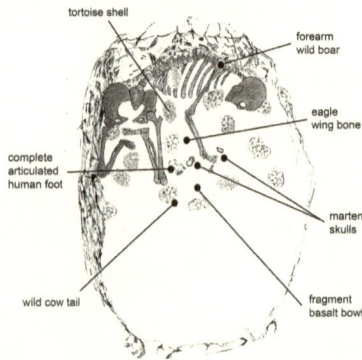

tortoise shell

forearm
wild boar

eagle
wing bone

complete
articulated
human foot

marten
skulls

wild cow tail

fragment
basalt bowl

on flowers.

She Was, She Looks

Good morphometrics, mDNA, and good dating
combined to nail it: a late
Pleistocene human skeleton was found
in Hoyo Negro, a submerged chamber of the
Sac Actun cave system in
the Yucatan Peninsula and

    this rare skull looks

intact and clean. It has
a long, high cranium, a projecting forehead;
a sharply angled occipital. The upper face
is short and broad relative to the neurocranium;
it has low, wide-set eye orbits, a broad nose.
Moderate alveolar prognathism, and lacking
the broad everted zygomatics
characteristic of people in the late Holocene and
contemporary Native Americana.

Here's more: she was gracile, small-statured,
and only 15 to 16 years of age
when she died; likely she fell
into a shallow pool
from an upper passage: perimortum fractures
of her pubic bones suggest it.

From one tooth
they determined this: she was

of Asian lineage, but of the type
specific to the Americas, meaning
genetically she descended
from those who came before her
across Beringia although now if fleshed out

she'd probably have
dark hair, big, wide-set eyes; her nose
would be low, wide and short; she'd have
a strong forehead, a severe
underbite. She'd look
          so much
like a negrito of the Philippines.
And note this, too, not reported. She died
in pain, in darkness,
without food,
frightened and alone.

News on the Pre-Clovis Front

As I drove home I saw
a young woman
standing on a street-corner in Oak Ridge; she had

a beautiful smile; she stroked
her hair as she talked with a friend,
all edges seem soft; it is as if
I'm living a dream. In Science

I find a strong thumbs-up
for the pre-Clovis folks—traces
of these ancient people are being found
all along the Pacific coast of
South America, and in Mexico, and now their bones
turn up mixed with stone flakes and a mastodon tusk in a
murky sink-hole beneath the Aucilla River
in Florida: the words

incontrovertible evidence
catch my eye. I learn
they did not survive
on big game exclusively; they ate
the little things: berries,
foxes, armadillos, like I eat
with my eyes
that hair-stroke and smile.

Kennewick Man

This day starts out complex
as a fine end-of-the-day red wine—who owns
Kennewick Man? A seal-eater,
it seems, but with a
leaf-shaped point bearing a serrated edge
in his hip and five broken ribs that failed
to heal properly.
Plus two dents in his skull—and his
9,000-year-old skeleton was found

more than a hundred miles
from the nearest seal. Did he walk
or row or paddle along the coast? If he did
probably he had
genes marking him as from Beringia.
They'll find that out I'm sure.
A 680-page book was needed
          just to summarize
what we know or think we know about him.

His anatomical features
differed from those of Native Americans
today and his relationship
to other ancient people is
uncertain. But back to now: what to do
with this day? It is

September 1. We know
we should rest, or work hard, or respect
those that work hard and today
in our back yard we're building
a future archeological delight: a terrace
where now the water flows. Flat rocks

pulled from a riverbed somewhere were brought here
on pallets by truck; each pallet weighs
more than a ton and the rocks
must be moved
        yes must be moved
by hand and wheelbarrow and stacked
        just so
and behind them on the uphill side
we'll put coarse gravel first, and top it
with good dirt to
level the field. No thick book is needed
for this work: we gain blisters,
we lose sweat.

Didja Know the Flood

An agrarian transition
from hunting wild sheep to herding them
began 10,200 years ago and unfolded
over centuries—it's been tracked
by analyzing piles of dung.

    Didja know

the level of pain perceived
by caged mice injected
with irritant in a foot
depends

on the gender of the person
injecting the mouse: it's less hurtful
when done by guy.

    Didja know

starfish now are dying
like mad along the west coast;
sunflower sea stars
are first, then giant pink sea stars, then

bat stars. They don't know why.

    Didja know

corn grows by vapor-pressure deficit,
neuronal activity promotes
adaptive myelination in the mammalian brain, and
interfacial effects
in iron-nickel hydroxide-platinum nanoparticles
enhance catalytic oxidation.

We Take, We Lose

We take
our first red-faced breath, we lose
our umbilical cord. Squinty-eyed, we peer first
uncomprehending the world
around our new little selves—things

touching the skin, the plump little digits
of our hands and feet; suckling
first microbes in milk from the mom,
then eating

billions and billions
of microbes on food, inhaling
millions more
with each breath until they become

life in us, too—streaming
the blood and the lymph, adhering
to other microbes in the gut, adhering
to the skin, the transparent
surface of the eye, to the inside

of the nose, the sinuses, and yes
even to the
rich waxy gold in the ears. We have
microbial roots in our cells;

they suck in glucose and pump out
carbon dioxide, ATPs, jiggly
high-energy electrons and these things

for life we lose: for life
we give up, give off, let go.

Found and Lost

Found,
this morning: a title, a

great title! Unfolding
the laws of star formation:
the density distri-
bution of molecular clouds.

It made me remember
long ago as a child when I read

Wynken, Blynken and Nod
up there, riding stars in a wooden shoe.

Now, a probability density function threshold
is described for star formation:
then, it was just a

child's idea,
drifting on a sea of stars.
On a starry night now

     we can ask if

Doggerland was Atlantis and if
both were washed away

in a great tsunami
eight thousand two hundred years ago
   caused by
the Storegga slide; such tsunami

would have been

a five-plus meters-high swelling
wall of oceanic water covering over

in Mesolithic times

bone spear-points, thousands of lives,
a great paradise.

When for a Day Now Old Fears

When for a day now
Mount Sinabung takes pause
we might pause too: the ash there
settles slow like casual snow
but people in the area may yet become
like those who stayed
too long in Pompeii—suf-
fering a suf-
focating hydrothermal pyroclastic flow.

Put for a moment a tight eye
to Pompeii: count backward
through types of calendars. The event

started near one o'clock
in the afternoon, August 24, 79 AD. Vesuvius,
from eyewitness account by Pliny the Younger,
initiated the disaster

with a great column
stabbing high into the air before
branching out—gas, ash, pumice, a
hundred thousand times
the thermal energy released
by the Hiroshima atomic bomb and

within two days
over a series of surges

some sixteen thousand people died.

What then? What about
the pyroclastic density currents

set up by the ash-column's slow collapse
Unveiled in technical discussions now
on the vector sum of magnetic fields
        which gave
investigators opportunity to infer
temperatures of roof tiles while
gases surrounding the city climbed
to incinerating temperatures. These things
were followed later by a meter more
of soft
white
ash.

Digging now through facts there's this:
the clothes they wore when carbonized
were heavier and warmer than one might expect
for August.

And olives and fresh fruit and vegetables
preserved in the shops were typical of October,
while fruits prevalent in August
were being sold instead in dried-fruit form.

And many wine-fermenting jars
had been sealed over—such sealing should have happened
near the end of October. And

        one coin

found in the purse of a woman curled under the ash
was commemorative, and likely had been minted
at the end of September.

Slippage in time

reconciled not by evidence but rather by
error in transitioning and translating
among calendar types, and

    when for a day now

we catch up at last we learn

in hippocampus cells
the enzyme HDAC2 patrols genes
silently where old memories are stored
under histones and turning HDAC2 off lets
the histones unfold so old memories
can get reworked and can re-
consolidate through time so old fears

attenuate, dissipate.

We Are Left There Now

In a cave
well above tide-line on the western tip
of Sumatra, one can face a stack-cake of history:
layers of sand and clamshells carried in
by tsunamis, with events
separated cleanly by layers of bat-dung:
eleven tsunamis before 2004,

the year of our Lord with some of the events
clustered and closer in time.

And note this too: at Santorini, north of Crete,
the African Plate is still
    sliding under
the Eurasian Plate. That situation caused hell
to break loose
thirty-six hundred years ago,
with the massive eruption of Thera:
that eruption destroyed
the Minoan civilization. What a
four-syllable mouthful, Santorini.

We are left there now

with rainwater, sweet
tomatoes that stain the jaws,
white eggplant, a dark
and delicious amber-orange wine.

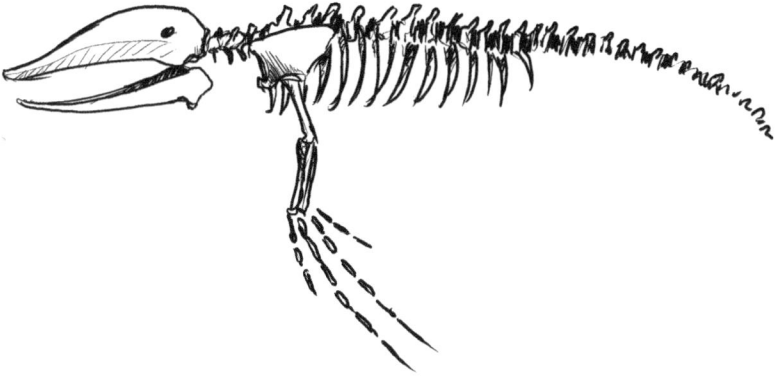

Whales of Cerro Bellena

A recent road development project
near the coast in the Atacama region of Chile
revealed fossil skeletons:
rorqual and sperm whales, seals,
predatory fishes and odd
species now extinct, such as
walrus-whales and aquatic sloths.

The gray-yellow bones
of more than forty whales are laid out
in four distinct time-horizons
so the mass strandings
must have occurred repeatedly

through time: so many species
and such large groups
of sea-animals collapsing in death in batches,
slick bodies rolling with waves to shore—why?

    They think

harmful algal blooms may have been to blame.
And yes, by good science, that's
the likely mechanism. But we can also ask

what do these collections of life-lost bones
mean for those of us now staring

stunned at the beautiful 3D-printed
likeness of a fossilized whale,
laser-scanned from the real thing?

Thirty thousand years from now some
creatures like us but different may look back
through time and dig
through layers of sand to find and study
dolls, timbers, house-hold goods and
radioactive bits of things
torn loose by the tsunami that hammered
the homes and reactors of Fukushima.

Again and again, it seems so clear:
things change; we lose, we gain,
and what we gain we lose.

We Move On

We've nearly forgotten

Yang and Lee's beautiful work
on the non-conservation of parity
under the weak force. But
    speak, memory:
what other things have so slipped?

From underground, for example:
here a purple crocus
pushes up from the Earth in spring,
and in Europe a large blue butterfly lands
delicately on wild thyme.

Or, while standing on stream-bank,
glance upstream—one
drifting yellowed tulip poplar leaf
    catches the eye, spins
around a rock on an eddy. Just like that,

things change:
over and over through time
our history gets burned
    slowly
into our DNA

and we move on.

Not Everything

Not everything is science—
or if it is, it's odd,

leaking and dribbling
things we don't yet know.

What, for example,
are we to make of the Antikythera mechanism,

or the sun's
coronal temperature problem,

or the Voynich manuscript,
or the Taos hum? Why

do we blush
and like to kiss?

Panspermia

Might have happened, and could be
in action now, who knows? Life,
spreading across the universe,
somehow making do in

hard space, despite
intense cold and cosmic rays;
affixed to or embedded in
asteroids, comets, bits of space-dust drifting
tens of millions of years, settling

    at last

here on Earth, and there
on Mars, and
here and there on other
life-hospitable
places we don't yet know.

What We Left

In his book *Paterson*, William Carlos Williams addressed
    just one place, one life—
        an entire book.

But why look so small, when such a grand
sweep of history is available?
Ten to twenty thousand years ago

our ancestors put painted hands
to the walls of caves; we scratched
notions in stone, we pushed
sharp sticks to dent the surface of wet clay; we hid
jewelry and coins, we left
arrowheads, flints, fibers, our very bones

for their great, great,
great granddaughters and grandsons to find.

New Flood in the Colorado Delta

When the flood happens
this year on schedule it will

flush salts from the dry beds;
    it will carry
the seeds of cottonwood and willows
to new sandbars,
    recharge
dry wetlands and you can bet
they'll scurry to use
LiDAR surveys
for high-resolution topography
and measure
changes in ground-water levels in shallow wells
and seed-germination success

    but
long ago now
Matevilya gave to the Mojave people
their names and their commandments;

and his son, Mastamho,
gave to the Mojave people
the river, and he taught them
about cottonwood and willows
and how to plant.

Fossil

I am holding a fossil: a fern-leaf,
about six inches in length, embedded in shale.
It was named
Alethopteris Pennsylvanicum
and it was found
in St. Clair County, Pennsylvania;
it was sold
by Hollow Mountain, a fossil dealer
in Louisville, Kentucky
        according to the business card
        taped to the fossil's back.

Fossils are a business now. At a web-site I find
a well-preserved 3-inch Green River fossil fish
for twenty-nine dollars—think about that!
You can buy
one thousand three hundred and ten
years or more of historical life

      for a penny.

A beautiful fossil crab with both claws
costs more than twice as much per year—
you can buy it, five hundred fifty-eight years

      for a penny.

And from a closed quarry near Pfalz, Germany,
something exceedingly rare: eight amphibians—
five Branchiosaurs and three Micromelerpeton.
Once they had soft bodies, but now
they're black ghosts pressed
rock-hard in stone. Studying
images of them on my computer now I know

I'll leave no footprints in sand or
handprints in red clay on cave walls; yet still I long
for mother-time to carry me forward, some-
how, one
slow lurch after another,
nestled on her back

How Quickly We Forget

The rise of armadillos in Tennessee,
the rise of feral hogs in Arkansas;
            the rise and fall
of fire ants and smallpox,
            the decline
of small mammals in northern Australia;
            the great big bang
of Krakatau;
            the disasters
of Minamata and Chernobyl;
the fireball and finger of smoke as

            in one quick flash

something
came down and made
Barringer Crater.

Lévy Walk

We know now how water vapor
behaves in air, and that lets us know now
   we're in
      Big Donkey Trouble
with respect to climate-change. Clouds
are less of a mystery: water vapor rises
15 kilometers, more or less, to make a cloud—
   or sometimes much less,
and then it makes
fewer clouds, which means

   Oh, boy, more sun, more heat
per unit of carbon dioxide—
four degrees or more, Celsius, by 2100,
they say and
   that's big.
And fast.

Yet unencumbered by future expectations,
we can look back and see

we walk the Lévy walk, for food, like honeybees.
This type of wandering path involves
a series of short movements in one area
and then a longer trek to another area—
   an action
not limited to searching for food.
We use, for example, a Lévy walk
while ambling in an amusement park,
and it describes
a pattern reported for urban development
and the formation of this poem.

The Legs, the Feet

My legs and feet
motor me up the sidewalk's little hills; I crest
each one and coast
going down. It is early
but still dark as night; it smells
like night and the humid air
muffles the small sounds of
insects busy making their small sounds.

My legs and feet
are special to me, having worked
so well for years. But now my left foot

is becoming
an aggravation: plantar
fasciitis, plus the gradual loss of
fatty tissues around the bones so

Wow! That first step
from bed in the morning
can be a doozey. Before
putting on socks and shoes I stretch
the toes, bending them up and back; I roll
the arch of the foot
over a bottle or on a golf ball;

it doesn't help much. So welcome, foot
problem, to the left-hand problem—
the thumb-base of which has become
arthritic. Such

decrepitude, I guess,
should be expected. Ascending,
I lose some steam, but going down,
I gain momentum.

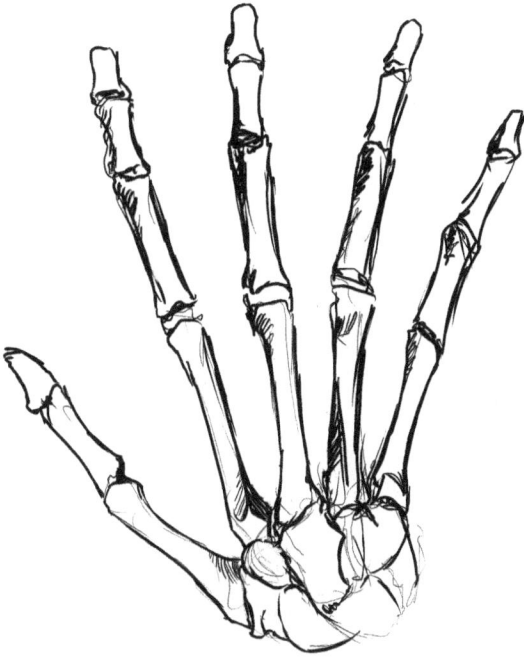

After a While You Learn

After a while you learn
the curves of the roads near home and
where the policeman hides his car
and the location of the patch
in early spring that sports
the clutch of daffodils.

After a while the hair grays and thins
and the joint
at the ball of the thumb
on the left hand hurts

    most of the time

and after a while you know
you've gained

    in advance

the knowledge in winter
of the day and time of night
when the moon's full face
will rise above the ridge.

How Is It?

We infer, we deduce, we
guess from many skinny facts. How is it
these abilities developed as
processes in our brains?

We hope, we plan, we
dare to live
day by day on skinny facts. How is it
these wants got hard-wired in our brains?

Angkor Wat

Having not been there before,
when I see it
for the first time on-line I must ask

    what happened,

making it empty now?
Drought, or disease? Was it
erosion of social structure from the flow
of religious ideas? A cunning stone face

peeking from a hole in the trunk of a tree
on site does not say

churning the cosmic
ocean of milk to release
the elixir of immortality released
a great poison, too;

it does not say

a blue-throated god, great serpents,
the stability
of the back of an immense tortoise;
Vaikuntha, Goloka, alive.

Like Shad Ascending

Heard
on National Public Radio
today: shad
are starting their annual run
up the Hudson River; the water temperature hit
fifty degrees so yes, they began

slipping up-river,
silver bodies, as William
Carlos Williams said
in the bitter world of spring, midway
between the surface and the mud.

On the radio
upon being interviewed about the run,
someone said,
when cooked, they taste
positively dreadful:
the fishiest, oiliest thing
        you can imagine
stuffed with bones,
covered with dirt. Yet

still, like us

they rise, year after year,
again and again, one
generation after another,
slipping upstream through time.

What's Happening

So, what's happening
today in the world? A quick check shows

polio in Pakistan, encephalitis in Egypt,
cholera in Haiti, and dengue fever

running rampant in Brazil, Mexico and India;
yellow fever

throughout sub-Saharan Africa;
Chikungunya

causing pain throughout the Caribbean and
pushing now into Florida;

Ebola killing and terrifying
people like mad in West Africa,

people dying
        everywhere and somewhere

(—Shhhh—! in China
likely, or India, based on the odds

of numbers), perfect in innocence, a
Christ-child, just born.

Deeper into Why

The moon they think
was created perhaps by impact
of a Mars-sized planet that struck the Earth
a glancing blow four and a half
billion years ago, that's long,

long before mankind oozed up
from primitive pre-man creatures;

long before dinosaurs roared in swamps

but note here the perhaps: unsettling data
on isotopic ratios of moon-rocks—
oxygen, titanium,
calcium, silicon, tungsten—
they should be like

something else, but they're more like
isotopic ratios of rocks on Earth.

So now what? Well, first look harder
at the isotopic ratios of the moon-rocks, make
more accurate measurements, dig

deeper into why.

The Language of Cross-Disciplinary Research

In our stream studies we found it
    so easy to slip
to short-hand terms and acronyms:
Campo for *Campostoma*, LMB for largemouth bass,
Spiro for *Spirogyra*, WMB for warmouth,
BG for bluegill—
just like that: short and slick,
    a new language
built and applied
for convenience and speed as we bumped
in a Suburban van
down rutted cut-back roads to streams

in new areas—outcrops,
limestone, slate, shale, conglomerate, flood-
ravaged gravel beds and pools,
great black basaltic boulders
slick with diatoms and bacterial slime,
bedrock, cobbles, pebbles, sand and silt;
tape measure, meter-stick, metric
rulers, we held

our binoculars up and ready; our
Rite-in-the-Rain yellow field notebooks

swelled as we learned.

Did Not

Sat down,
got up, adjusted the chair,
sat down again: checked, what things
are flying through the head
this morning, causing

upset or roil; what things
do I need to attend to
now, in order to

establish order, create
a pebbly path or a rabbit-worn
hop-trail
past the garden toward the bush

and
what things, seriously
need attention
if I'm to write
anything worth a tinker's dam

    (look it up)

I'm only here

such a brief while to point
to that thing or another, and tip
my hat (if I had one) to the long-ago
trudge and sweat of
others before me: I've no special
wisdom to share; my life

is not great or big, did not
stop and turn aside, astonished.

Whimsicality

We're astonished
daily by what we learn, it really is
astonishing: something like
the thing we feel with atonement—a
reconciliation of truth
or agreement
with what we see as fact and

there, in fossil form, an ancestor of the modern
velvet worm is now
revealed: what's that,
you say, a velvet worm? A good
ask—I say, I look it up: it's not
a velvet ant, it's more like
a small worm with clawed legs, or a caterpillar,
one of two groups being found
now in what once was

Gondwana: well before people—odd but true
and here the bifurcation leaps to past
by virtue of the fossil: a 1.4-inch
little thing, a shadow-cast
creature in Burgess Shale, found

forty years ago and clocking in
at 505 million years of age, and

that little sea-floor creature they named
Hallucigenia sparsa, not knowing
leg from spine or head from tail.

His Obituary in Advance

Here are some things
he said were successes, put
in random order so as to protect
others in the world:
in the glory-days of high school
a half-mile best time of 2:03.1, and a
mile-time best of 4:37.2. Much later he won
a wonderful woman who became
his beautiful wife. They had
a blended family, comprised
of his three children from an earlier marriage
plus two of hers: and each one of them knows
exactly who they are. A little, not enough,

success on the science side; a little, not enough
success on the writing side—a 2013 inductee
into the East Tennessee Writers' Hall of Fame
    for poetry.

He lived
a fairly long time—a dozen years or so
in Indiana; two years
in the Peace Corps in Ghana, West Africa;
more than 25 years
in East Tennessee; several years each
in Oklahoma, Michigan, and a few more than that
in Arizona. He is survived, he hopes,

by his wife, his three children and her two, and
by his two brothers and two sisters,
and his mother, alive now, but probably not for long.

He said may they all share
the remarkable blessings of this earth.
He said
he tried to do good.

About the Author

Arthur Stewart was born in Michigan City, Indiana, and lived initially in what is now the Indiana Dunes National Park. He spent time in Arizona, where he studied biology and chemistry at Northern Arizona University. After two years in the Peace Corps, he then earned his PhD in aquatic ecology at Michigan State University. His postdoctoral research was done at Oak Ridge National Laboratory (ORNL), where his research focused on the toxicity of coal oil and shale oil. Then he taught and did stream ecology studies at the University of Oklahoma before returning to ORNL, where he worked as an ecotoxicologist, group leader and senior scientist for seventeen years. To pursue his developing interests in improving science education, Art earned a MS Ed at the University of Tennessee – Knoxville, and became a project manager for Oak Ridge Associated Universities (ORAU). Within ORAU, he manages various science education and workforce development programs. He lives in Lenoir City, not far from the hubbub of Knoxville, Tennessee and writes whenever possible.

Notes

Science-Flavored Poetry (Page 3)
Oliver, M. 1994. A Poetry Handbook. Harcourt, Inc., San Diego, CA. 130 p.

From Where We Came (Page 4)
This poem was published originally in Cultural Studies of Science Education, DOI 10.1007/s11422-014-9624-x.

Long, Long Ago (Page 8)
Nance, D.R. et al. 2006. Acatlán Complex, southern Mexico: Record spanning the assembly and breakup of Pangea. Geology 34(10):857-860; doi: 10.1130/G22642.1

The First, Second and Third Worlds According to the Hopi (Page 9)
Waters, F. 1977. Book of the Hopi. Penguin Books, New York, NY. 345 p.

Denisovans (Page 14)
Meyer, M.; Kircher, Martin; Gansauge, Marie-Theres; et al. A high-coverage genome sequence from an Archaic Denisovan individual. Science 12 October 2012: 222-226.

To the Time of Long Ago (Page 15)
Verweij, W.; Spelt, C.; Di Sansebastiano, G-P.; Vermeer, J.; et al. 2008. An H+ P-ATPase on the tonoplast determines vacuolar pH and flower colour.

Nature Cell Biology 10(12):1456-1462.
http://www.nimr.org/research/cardiac.html (accessed 20 September 2014)

Lesson on Ancestral Aboriginal Australians (Page 17)
Rasmussen, M.; Guo, X.; Wang, Y.; et al. 2011. An Aboriginal Australian genome reveals separate human dispersals into Asia. Science (7 October 2011): 94-98.

End of the African Humid Period (Page 21)
Bard, E. 2013. Out of the African Humid Period. Science (15 November 2013): 342(6160):808-809.

Ice-Man Ötzi (Page 22)

Müller, W.;Fricke, H.; Halliday, A.N.; McCulloch, M.T.; Wartho, J.-A. 2003.

Origin and migration of the Alpine iceman. Science (31 October 2003) 302: 862-866. http://www.iceman.it/en/node/226, accessed 19 September 2014.

Like and Unlike (Page 24)

Gibbons, A. 2011. Skeletons present an exquisite paleo-puzzle. Science (9 September 2011) 333:1370-1372.

Bluestone Rocks of Stonehenge (Page 25)

Bevins, R.E.; Ixer, R.A.; Pearce, N.J.G. 2014. Carn Goedog is the likely major source of Stonehenge doleritic bluestones: evidence based on compatible element geochemistry and Principal Component Analysis. Journal of Archaeological Science 42(February 2014): 179–193.

Yes and No for the Pigs (Page 26)

Balter, M. 2014. Life and death at Stonehenge. Science (3 January 2014) 343:20-23.

Peccary-Chase (Page 28)

Gibbons, A. 2014. New sites bring the earliest Americans out of the shadows. Science (9 May 2014) 344: 567-568.

Closer to Home (Page 29)

Akpan, N. 2014. Lasers unearth lost 'agropolis' of New England. Science news (10 January 2014; 5:45 pm).

Easter Island (Page 30)

http://www.npr.org/blogs/krulwich/2013/12/09/249728994/what-happened-on-easter-island-a-new-even-scarier-scenario (accessed 20 September 2014).

http://scienceblogs.com/notrocketscience/2009/07/08/rapamycin-the-easter-island-drug-that-extends-lifespan-of/ (accessed 20 September 2014).

Where, When and How (Page 31)

White, T.D. 1980. Evolutionary implications of Pliocene hominid footprints. Science (11 April 1980) 208:175-176.

Gone Before it Was (Page 32)

Lordkipanidze, D.; Ponce de León, M.S.; Margvelashvili, A.; Rak, Y.; Rightmire, G.P.; Vekua, A.; Zollikofer, C.P.E. 2013. A complete skull from Dmanisi, Georgia, and the evolutionary biology of early Homo. Science (18 October 2013) 342:326-331.

Chinchorros (Page 37)

Marquet, .PA.; Santoro, C.M.; Latorre, C.; et al. 2012. Emergence of social complexity among coastal hunter-gatherers in the Atacama Desert of northern Chile. Proceedings of the National Academies of Sciences 109:14754-14760.

Natufians (Page 38)

Power, R.C.; Rosen, A.M.; Nadel, D.; 2014. The economic and ritual utilization of plants at the Raqefet Cave Natufian site: The evidence from phytoliths. Journal of Anthropological Archaeology 33:49-65.

Robert Power, of the Max Planck Institute for Evolutionary Anthropology, kindly reviewed an earlier version of Natufians for technical accuracy.

The illustration with the poem is by P. Groszman, Jerusalem, Israel, and was published first in Grosman, L.; N.D. Munro; A. Belfer-Cohen. 2008. A 12,000-year-old Shaman burial from the southern Levant (Israel). Proceedings of the National Academy of Sciences of the United States of America 105(46):17665-17669.

She Was, She Looks (Page 39)

Chatters, J.C. et al. 2014. Late Pleistocene human skeleton and mtDNA link Paleoamericans and modern Native Americans. Science (16 May 2014) 344:750-754.

James Chatters, of Washington University, kindly reviewed an earlier version of She Was, She Looks for technical accuracy.

News on the Pre-Clovis Front (Page 41)

Gibbons, A. 2014. New sites bring the earliest Americans out of the shadows. Science (9 May 2014) 344:567-568.

James Chatters, of Washington University, kindly reviewed an earlier version of News on the Pre-Clovis Front for technical accuracy.

Kennewick Man (Page 42)

Owsley, D.W. and R.L. Jantz (eds) 2014. Kennewick Man: The Scientific Investigation of an Ancient American Skeleton. Texas A&M University Press, College Station, TX. 680 p.

Didja Know the Flood (Page 44)

Grimm, D. 2014. Male scent may compromise biomedical studies. Science (2 May 2014) 344:461

Ort, D.R.; Long, S.P. 2014. Limits on yields in the Corn Belt. Science (2 May 2014) 344:484-485.

Gibson, E.; et al. 2014. Neuronal activity promotes oligodendrogenesis and adaptive myelination in the mammalian brain. Science (2 May 2014) 344:487.

Stokstad, E. 2014. Death of the stars. Science (2 May 2014) 344:464-467.

Chen, G.; et al. 2014. Interfacial effects in iron-nickel hydroxide-platinum nanoparticles enhance catalytic oxidation. Science (2 May 2014) 344:495-499.

We Take, We Lose (Page 45)

Couzin-Frankel, J. 2014. A gut microbe that stops food allergies. http://news.sciencemag.org/biology/2014/08/gut-microbe-stops-food-allergies

Found and Lost (Page 46)

Kainulainen, J.; Federrath, C.; Henning, T. 2014. 2014. Unfolding the laws of star formation: the density distribution of molecular clouds. Science (11 April 2014) 344:183-185.

Weninger, B., Schulting, R., Bradtmoeller, M., Clare, L., Collard, M., Edinborough, K., Hilpert, J., Joeris, O., Niekus, M., Rohling, E., et al. 2008. The catastrophic final flooding of Doggerland by the Storegga Slide tsunami. Documenta Praehistorica 35:1-24.

When for a Day Now Old Fears (Page 48)

Zanella, E.; Gurioli, L.; Pareschi, M.T.; Lanza, R. 2007. Influences of urban fabric on pyroclastic density currents at Pompeii (Italy): 2. Temperature of the deposits and hazard implications. Journal of Geophysical Research – Solid Earth 112 (B5): B05214, DOI: 10.1029/2006JB004775.

Miller, G. 2012. How are memories retrieved? Science (5 October 2012) 338:30-31.

We Are Left There Now (Page 51)
http://www.nrf.gov.sg/scientific-advances/science-spotlight/a-treasure-trove-of-ancient-tsunamis (accessed 20 September 2014).

http://en.wikipedia.org/wiki/Santorini (accessed 20 September 2014).

Whales of Cerro Bellena (Page 55)
Pyenson, N.D.; Gutstein, C.S.; Parham, J.F.; et al. 2014. Repeated mass strandings of Miocene marine mammals from Atacama Region of Chile point to sudden death at sea. Proceedings of the Royal Society, B-Biological Sciences. 281(1781), 20133316.

http://youtube/nMHofHrxuGQ (accessed 20 September 2014).

Nicholas Pyenson, of the Smithsonian National Museum of Natural History, kindly reviewed an early draft of Whales of Cerro Bellena for technical accuracy.

We Move On (Page 57)
http://physics.nist.gov/GenInt/Parity/expt.html (accessed 20 September 2014).

Nabokov, V. 1966. Speak, Memory: An Autobiography Revisited. G.P. Putnam's Sons, New York, NY. 316 p.

Not Everything (Page 58)
Freeth, T. 2014. Eclipse prediction on the ancient Greek astronomical calculating machine known as the Antikythera Mechanism. PLoS One 2014, 9(7): e103275.

http://www.astronomy.com/news/2013/10/astronomers-find-clues-to-decades-long-coronal-heating-mystery (accessed 20 September 2014).

http://www.livescience.com/43519-taos-hum.html (accessed 20 September 2014).

Amancio, D.R.; Altmann, E.G.; Rybski, D.; Oliveira, O.N., Jr.; Costa, L.d.F. 2013. Probing the statistical properties of unknown texts: application to the Voynich Manuscript. PLoS One 2013, 8(7): e67310.

Panspermia (Page 59)
Mendonca, M.d.S., Jr. 2014. Spatial ecology goes to space: Metabiospheres. Icarus 233:348-351.

What We Left (Page 60)
Williams, W.C. (Revised edition prepared by C. MacGowan). 1992. Paterson. New Directions, New York, NY. 311 p.

New Flood in the Colorado Delta (Page 61)
Stokstad, E. 2014. U.S. and Mexico unleash a flood into Colorado delta. Science (21 March 2014) 343:1301.

Ramírez-Hernández, J.; Hinojosa-Huerta, O.; Peregrina-Llanes, M.; Calvo-Fonseca, A.; Carrera-Villa, E. 2013. Groundwater responses to controlled water releases in the limitrophe region of the Colorado River: implications for management and restoration. Ecological Engineering 59:93-103.

http://www.mojaveindian.com/creation.htm (accessed 20 September 2014)

Fossil (Page 62)
http://www.fossilrealm.com/products/large-rare-prehistoric-amphibian-sclerocephalus-from-pfalz-germany (accessed 20 September 2014).

Lévy Walk (Page 65)
Raichlen, D.A.; Wood, B.M.; Gordon, A.D.; Mabulla, A.Z.P.; Marlowe, F.W.; Pontzer, H. 2014. Evidence of Lévy walk foraging patterns in human hunter-gatherers. Proceedings of the National Academy of Sciences. 111 (2):728-733. DOI: 10.1073/pnas.1318616111.

The Legs, the Feet (Page 66)
http://www.mayoclinic.org/diseases-conditions/plantar-fasciitis/basics/definition/con- 20025664

After a While You Learn (Page 71)
http://en.wikipedia.org/wiki/Winter_solstice

How Is It? (Page 72)

Alink, A.; Schwiedrzik, C.M.; Kohler, A.; Singer, W.; Muckli, L. 2010. Stimulus predictability reduces responses in primary visual cortex. J. Neurosciences 30(8):2960-2966.

Nortmann, N.; Rekauzke, S.; Onat, S.; König, P.; Jancke, D. 2013. Primary visual cortex represents the difference between past and present, cerebral cortex, DOI: 10.1093/cercor/bht318

Angkor Wat (Page 73)

Evans, D.H. et al. 2013. Uncovering archaeological landscapes at Angkor using lidar. Proceedings of the National Academy of Sciences. 110 (31):12595-12600. DOI: 10.1073/pnas.1306539110

Anonymous. 2012. Ancient canals transported building blocks to Angkor Wat. Science (19 October 2012) 338 (6105): 312. DOI: 10.1126/science.338.6105.312

Like Shad Ascending (Page 74)

Williams, W.C. 1976. The Bitter World of Spring, p. 164, in Selected Poems. Ed. by C. Tomlinson. New Directions, New York, NY. 302 p.

Deeper into Why (Page 76)

Day, J.M.D.; Moynier, F. 2014. Evaporative fractionation of volatile stable isotopes and their bearing on the origin of the Moon. Philosophical Transactions of the Royal Society A – Mathematical Physical and Engineering Sciences. 372 (2024), Article Number: 20130259. DOI: 10.1098/rsta.2013.0259

The Language of Cross-Disciplinary Research (Page 77)

Power, M.E.; Matthews, W.J.; Stewart, A.J. 1985. Grazing minnows, piscivorous bass, and stream algae: dynamics of a strong interaction. Ecology 66:1448-1456.

Did Not (Page 78)

Ammons, A.R. 1981. Easter Morning, p. 19, in A Coast of Trees. W.W. Norton & Company, New York, NY. 52 p.

Whimsicality (Page 79)

Smith, M.R.; Ortega-Hernández, J. 2014. *Hallucigenia*'s onychophoran-like claws and the case for Tactopoda. Nature. doi:10.1038/nature13576.

His Obituary in Advance (Page 80) Stewart, A.J. 2005. Bushido: Ghanaian Story. P
· 6, in The Virtues of Rei and Makoto. Celtic Cat Publishing, Knoxville, TN.
80 p. (CelticCatPublishing.com)